Four Powerful Words in Apologetics

Other books by Patrick Nurre:

Rocks and Minerals for Little Eyes (PreK-3)
Fossils and Dinosaurs for Little Eyes (PreK-3)
Volcanoes for Little Eyes (PreK-3)
Geology for Kids (and Geology Journal) (3-6)
Rock Identification Made Easy (3-12)
Rock Identification Field Guide
Fossil Identification Made Easy (3-12)
Fossil Identification Field Guide
Mineral Identification Made Easy (5-12)
Bedrock Geology (high school)
Rocks and Minerals: The Stuff of the Earth (high school)
Volcanoes, Volcanic Rocks and Earthquakes (high school)
The Geology of Yellowstone – A Biblical Guide
Genesis Rock Solid – A Biblical View of Geology
Fossils, Dinosaurs and Cave Men (high school)
Geology and the Hawaiian Islands (high school)
Geology and Our National Parks (high school)

These are all also available with sample rock, mineral, and fossil kits at NorthwestRockAndFossil.com.

Four Powerful Words in Apologetics

Patrick Nurre

Four Powerful Words in Apologetics
Published by Northwest Treasures
Bothell, Washington
425-488-6848
NorthwestRockAndFossil.com
northwestexpedition@msn.com
Copyright 2020 by Patrick Nurre.
All rights reserved.
Cover designed by Heather Hall.

Printed in the United States of America. No part of this book may be reproduced in any manner whatsoever without written permission except in the case of brief quotations embodied in critical articles and reviews.

Scripture quotations taken from the New American Standard Bible®
Copyright © 1960, 1962, 1963, 1968, 1971, 1972, 1973,
1975, 1977, 1995 by The Lockman Foundation
Used by permission. (www.Lockman.org)

Contents

6   Introduction

8   Knowing the Scriptures in the Secular Scientific Age

14   Clarifying the Conflict between Science and the Bible – Four Powerful Words

Four Practical Exercises in Learning the Four Powerful Words

24   Exercise 1 - The Genealogies and Chronologies of Genesis: Are They Accurate and Reliable?

35   Exercise 2 - Evolutionary Gaps in the Fossil Record: How Serious Are They?

47   Exercise 3 – Dinosaur-to-Bird Evolution, the Story that Never Seems to Die

56   Exercise 4 – Time and Chronology in the Secular Scientific Age

66   Credits and References

Introduction

I believe that there are four very powerful words that if learned and used, can help you to truly understand the conflicts involved in modern apologetics, and to acquire enough understanding to answer many of the intimidating questions used by modern scientists that intimidate those who trust the Scriptures.

As Christians engage in apologetics for the faith, we often put the cart before the horse. In other words, we often make our view of Scripture subservient to the authority of man. When our book of Genesis is assaulted by scientists claiming that most of Genesis is non-scientific and consequently just myth, what do we do with that? Here are three typical responses:

1. We think that because these scientists are professionals, they have mastered their field and surely should know what they are talking about.
2. We think that we are being unreasonable and inflexible and tend to think that we should probably reinterpret portions of Genesis to be more in line with what scientists say.
3. We become silenced into doubt and unbelief.

We must remember, as Christians, that our authority is Scripture, not man, making the authority of Scripture our starting point in all things. My desire is for you to be thoroughly equipped in the Scripture as those words which have been God-breathed, and to gain a confidence that you can use the Scripture to tear down false ideas and false concepts put forward as truth by the science community.

Four Powerful Words is written to help you not lose your way in the challenge of apologetics and communicating the truth of the gospel. If we can remember to use this tool not only to help our own faith but also to help others who are being confused by the tangled web of words and their meanings in modern science, we will do a better job of challenging people to think about what they believe and why.

Most of my references will be in the context of geology because it has universally affected every area of science beginning in the 1700s. Although many of the first scientists were Bible-believing, it was the field of modern geology, beginning in the late 1700s, which turned the scientific culture of the day to its totally atheistic framework of today.

I trust that the information in this book will help you gain more confidence in the Scriptures as the word of God and that your faith becomes stronger as a result. Thank you for taking the time to read this valuable book.

Please note that the intent of this short book is not to tackle all the particular questions and issues you might have. It is only to help you sort out the truth of what is science and what is not.

At the end of each chapter, I include a list of questions. Some of the answers to these questions can be found within the text of the lessons. Others are for discussion.

Patrick Nurre
May 20, 2020

Knowing the Scriptures in the Secular Scientific Age

What is apologetics?
The word apologetics comes from the Greek word, *apologia*, from which we get our English word, *apology*. But that is not what it means. Apology as is used in our modern language means to be sorry for or to feel bad for what I did or said. That is not the Biblical meaning. As the word is used in Scripture, it has several meanings, including:

- A defense
- An answer
- Making things clear or as we sometimes say, "clearing the air"

The modern church uses this word to mean the discipline of marshaling a whole host of evidences to support the Christian faith, such as proper theology, books and commentaries, quotes by famous people of science, rigorous training, and lots of evidence. But this is too narrow. It can include these things, but the word *apologia* is broader.

When we think of apologetics and those who are apologists, we typically think of somebody, a specialist, who has studied and memorized facts and figures and can arrange them in an orderly manner to convince others that the Christian faith is true. This could be included, but it falls far short, as it leaves most of the Christian community out of involvement.

Think about it: How many experts in apologetics do you personally know? How many would you find in the

Christian community? Not many. Most of us don't have the time or the energy to devote to that. We feel more comfortable leaving it to others.

Apologetics and the Scriptures

But there are a number of Scriptures, which indicate that apologetics is for every believer. I have listed a few of these passages below. Take some time to familiarize yourself with their content:

- 1 Peter 3:13-16
- 2 Corinthians 10:3-5
- 2 Corinthians 7:11
- 2 Timothy 4:16-17
- Jude 3
- Romans chapters 1-11 (Paul's defense of the faith)

You probably noticed that some of these passages contain the word *apologia*, some don't. But the concept behind apologetics in these Scriptures is that Christians have a faith and a hope that the world needs. Consequently, all Christians need to be engaged in the apologetics that influences faith. As Hebrews 11:6 tells us, *without faith it is impossible to please God.*

All Christians must stay involved and engaged with the world around them for one main reason – to witness to that faith, and to explain that faith and hope. It does involve learning some evidence. That is true. But it also involves learning how to identify preconceived ideas, incorrect assumptions and speculations raised up against the knowledge of God, and to defeat them. (2 Corinthians 10:5) This is apologetics.

When it comes right down to it, winning an argument really does not depend on you, but on God's work in the hearer. After all, the hearer is at war with God. But you can work with God in this process by speaking to the conscience that God gave to each man and woman.

Apologetics and science

Apologetics is no more applicable than to the subject of science. Science, as broadly taught today, is more than the exploration of physical properties, chemicals, animals, and plants – the world we live in. It has, by and large, embraced a worldview that is not about science. This worldview excludes God and the Scriptures, in favor of naturalism and the exaltation of man. But it wasn't always this way. Where did this type of reasoning come from? We need to go back and take a look at the Enlightenment for part of the answer to that question.

Science and the Enlightenment

During the Enlightenment, (17th-19 centuries) the idea of compartmentalization began to take shape in Western Civilization. The generally accepted view was that we should leave matters of science to the scientists and matters of religion should be left to the church. This was a subtle shift that taught that only qualified scientists could interpret matters of science and nature. But by creating this division, the Scriptural explanation of our world that God created was undermined, and, therefore, faith in the Scriptures. If the Bible was not to be trusted when it spoke of the Creation and the Flood, then why would it be trusted when it spoke of other issues?

This lack of trust in the Scriptures has led many, inevitably, to relativism: that is, a lack of moral accountability to the

Scriptures. This has had its outworking in many ways, but especially in our lifetime, it has included redefining marriage, and confusion of sexual identity. And when you remove the authority of the Scriptures, it also removes a sense of purpose for our lives.

Knowing our Scriptures

So, we must know our Scriptures. Apologetics involves more than winning an argument by sheer weight of the evidence. It means that we clarify the issues and set the record straight. Regardless of how much so-called evidence could be marshalled against the Scripture, we as Christians must believe the Scripture and therefore must master its content and an ability to communicate it clearly. That is a Christian's domain. The Christian, of all people, has been enlightened by the Holy Spirit. And the Christian, of all people, should be able to discern that which is contrary to God's word and how to set the record straight.

I would suggest that the first step in developing apologetics is to know and master the Scriptures. I have personally discovered that many Christians do not know the Scriptures, nor have they wrestled through understanding them. We must become convinced that every word in Genesis is from God. Paul expresses this in 2 Timothy 3:16, "All scripture is inspired by God...." (*"Inspired by God"* is often translated, *"God-breathed."*) Are we personally convinced of this? This must be resolved before we can proceed to other areas of apologetics. It does no good to have all the answers about the resurrection of Jesus Christ from the dead, if we are wavering when it comes to what Genesis says. Sooner or later it will dawn on us that the God who spoke the gospels, also spoke Genesis. And it will be at this point that we will either agree that all of Scripture must have been spoken by

God, or we will eventually doubt all of the Scripture as directly breathed by God.

Questions:

1. What does the word, *apologia*, mean?

2. What does the activity of apologetics include?

3. Give a short descriptive sentence summarizing each of the passages listed as referring to apologetics.

   - 1 Peter 3:13-16
   - 2 Corinthians 10:3-5
   - 2 Corinthians 7:11
   - 2 Timothy 4:16-17
   - Jude 3
   - Romans chapters 1-11 (Paul's defense of the faith)

4. What is the first step in understanding apologetics?

5. Why should that be the first step?

Clarifying the Conflict between Science and the Bible – Four Powerful Words

Why can't we just all get along and live and let live? Everyone wants peace. The cornerstone of the United Nations heralds, "They will hammer their swords into plowshares and their spears into pruning hooks. Nation will not lift up sword against nation, and never again will they learn war." Isaiah 2:4. This was the noble goal of the United Nations when it was first created at the end of World War II. But, unfortunately, that has never happened and today seems even more of a remote possibility. There is never-ending conflict in our world and the only one who can put a stop to it has not come yet. Of course, we believers have just begun to experience the promised peace in Isaiah 2 because our Savior has come the first time to bear our sins and to fulfill the Scripture. We have been restored to a right relationship with God through Christ. But that is not yet our complete salvation. Complete salvation includes a new heaven and new Earth. And therein lies the conflict. The conflict between science and the Bible is not over scientific facts. It is over a view of the earth, its origin, its destiny, and its present state of existence.

One of the greatest misconceptions and deceptions about modern geology and much of modern science is the thinking that the conflict between science and the Bible is between hard scientific facts and illiterate, myth-believing Christians. That is such a distortion that we must give time to exposing this way of thinking and engaging peoples' attention about this.

After our knowledge of Scripture, the second area of apologetics we must become sharp in, is to learn and clear up meanings of four powerful words: **science**, **history**, **philosophy**, and **consensus**. Knowing and using these words in your discussions with those who insist that the Bible is just myth, will help keep your discussions on point. It will also make the unbeliever defend his faith, instead of a believer always being on the defensive. So, let's take a look at the four words that can help you to stand in your faith.

1. Science

The word *science* means knowledge. It is limited in its ability to understand things outside of the physical world. It deals with what can be known by the primary senses. Science deals with what can be:

- observed
- tested
- repeated

Let's examine this in light of the Big Bang, which is largely believed to be how our universe came to be.

What aspect of the Big Bang idea of the origin of the universe fits this definition? Not much of it at all. What aspect of the Biblical idea of creation can be observed, tested, and repeated? Well, not much of it, either. For instance, both speak of the stars and the sun, moon, and planets. But their origin is beyond the definition of science. That would involve needing eyewitnesses, since origins deals with historical events. We can see distant stars, planets, and comets whizzing by, but these things don't explain their origins, the how of how they came to be.

Many people have inordinately trusted in science to give them the answers to all of life's questions, especially the question of origins of our earth and of man. Many of the published statements of science that flood the internet these days about the origin of the universe, Earth, and life are things that are non-scientific. They are ideas couched in scientific terminology. The only reason the Big Bang is allowed to be discussed as a matter of science, is it is believed to be scientific, and because the only alternative to it is the special creation spoken of in Genesis. And God forbid that we should allow that.

What is your worldview?

We must remember that science is limited to what can be observed, tested, and repeated. Because of these limitations of science to explain all things, it is impossible for it to carry the authority to explain definitively the origin of the earth. That is not a matter of science. It is a matter of *belief*. Secular scientists *believe* certain things about our origins not because of evidence, but because of a worldview that excludes God. This is no different than the Christian who *believes* in a supernatural creation. We *believe* certain things about the earth because of a worldview we have chosen. And we must not forget that this worldview was the same as what Jesus Christ embraced!

Most secular scientists will not admit that they have a worldview. This is an area that they are blind to in their own biases. To admit to a worldview would mean that they are bringing non-scientific ideas to their inquiries. This would be contrary to scientific pursuits, so they will not concede that they have a worldview that influences the way they approach their field of science.

But one scientist did admit to this worldview in a startling quote:

> Evolutionists ... have a prior commitment, a commitment to materialism. It is not that the methods and institutions of science somehow compel us to accept a material explanation of the phenomenal world, but, on the contrary, that we are forced by our a priori adherence to material causes to create an apparatus of investigation and a set of concepts that produce material explanations, no matter how counter-intuitive, no matter how mystifying to the uninitiated. Moreover, that materialism is absolute, for we cannot allow a Divine Foot in the door. *Richard Lewontin*[1]

Scientific inquiry is not always the driver in the pursuit of knowledge. It is sometimes a commitment to a way of thinking, a belief. This belief will get in the way of interpreting discoveries. It is inevitable. Science should stick with what can be **observed, tested,** and **repeated.** There is no disagreement between secular scientists and Christian scientists on true science. Almost no one disagrees with actual science. But we must have agreement on the meaning of the word *science*, or we will simply be arguing. Arguments usually lead to angry tempers and words. As science means knowledge gained by observation, testing and repetition of those tests and observations, it is not a way of thinking! Clearing the air about this will help immensely in our conversations about origins! That brings us to the next word that needs a definition.

2. History

The word *history* deals with one-time, unique events that happened in the past. This past could be many years ago or just yesterday. But eyewitnesses and written documents to the actual events that have taken place in the past are what are needed to verify these historical events. The Bible has plenty of those! It is precisely here that the secular view of the origin of the universe and of the earth falls outside of history because it has no written records nor eyewitness accounts. It relies on personal interpretation of what is observed in the present to explain the past. Huge leaps of interpretation and imagination defined in scientific terminology are then employed to teach what are called facts of science.

So, a fair question to ask would be, "What does the Bible have to say about the origin and the development of the earth?" Although the accounts given in the Bible do not give us as much detail as we would like, the various accounts do give us *framework* in which to interpret the various landforms and geology we see in the earth.

For example, in Genesis 1:1, we are told that at the very beginning of what God had created, He created the space and the earth. That implies volumes about what happened in the beginning. It describes both a geological and a historical event.

Another example is found in Genesis 7:11. We are told that on the first day of the Flood, all the fountains of the great deep burst open. That statement also describes both a historical event and a geological event. While all the details are not given, a general idea of what happened on the first day of the global flood can be outlined. It is very easy to do. The only question is whether these accounts are historical and true. If they are not historical and true, then the whole of the Bible must also be suspect. And this would include the Biblical teaching about a Messiah and Savior. This brings us to the next word that needs to be defined.

3. Philosophy

Philosophy means, *love of wisdom* and it is the study of ideas. It involves interpretation. This is the starting point for where secular science begins to mix things up. Think about those late nights with friends, discussing the great questions of life. How did we get here? What is my purpose? What happens when we die? Everybody has an idea. But these ideas are just our beliefs. They are not science. They may seek to explain what I know, but these ideas remain outside the realm of science, and are firmly planted in the realm of philosophy.

Much of modern geology is philosophy or interpretation. For instance, someone might look at a rock and determine its mineral makeup. That would be science. But when I look at that rock and determine that it proves an ancient Earth of millions of years, I have entered the area of philosophy. And this brings us to the very misunderstood area of radiometric dating.

Radiometric dating

Radiometric dating crosses paths with three of the words that we have gone over: science, history, and philosophy. Radiometric dating is based on the **science** of radioactivity. Radioactivity can be observed and measured – in the present. But radiometric dating is an interpretive application of radioactivity. Why? Because it involves assumptions about the behavior of radioactivity in the ancient past. That pushes it into the realm of **history**. In studying radiometric dating, scientists make assumptions about the behavior of present radioactivity. Has it always behaved the same way that we observe in the present? But radioactivity was only discovered in 1897. Before that, we have no records.

Because we do not have a historical document throughout the long ages of Earth science proposed by modern geology, we really lack a lot of information. If the Creation and the Genesis flood were actual historical events, how could these events have changed our assumptions about the past behavior of radioactivity? In particular, might the Flood have had some effect on radioactivity that is not being accounted for in current dating ideas? This is an intriguing question that many scientists wonder about. But we just do not know and neither do we have the documentation to make that determination. In modern science, however, radiometric dating has settled itself in the halls of science and history, rather than where it rightfully belongs, in the school of **philosophy**.

If we can listen closely to statements that we hear and try to discern which things in the statements are science, history and philosophy, we can not only keep ourselves from getting confused, but can also help others sort some things out. We can help them "clear the air."

4. *Consensus*

Now we come to the fourth powerful word – *consensus*. Consensus is group agreement. It is often group agreement by very powerful people. And it can be regardless of the facts. Consensus is what largely motivates a silence among many scientists who disagree with the majority. They are not free to share their disagreements because often this has resulted in loss of jobs and reputations.

Evolution is an example of **consensus**. Evolution has never been proven to be true. Yet, it would seem that most scientists now believe that evolution is true science. That is considered a consensus of scientists. How could you disagree? In fact, I am convinced that this is why there is such a huge division in the church today over the age of the earth, because we fail to distinguish between actual science and consensus. Many Christians find it absurd to proclaim what the Bible clearly states about the age of the earth when it looks like the science community by and large says that this issue has been resolved by science. Just because there is a consensus does not mean that it is right or that it is a fact of science. We must teach ourselves to practice these principles and to teach them to others.

A good example of consensus can be found in the history of biogenesis. As recently as the late 1700s-early 1800s, it was believed by many that life could spring from non-life: this idea was called spontaneous generation. This belief was held tightly by many scientists without regard to proof of its veracity. It was Louis Pasteur that ultimately proved this idea false in 1864, by **observation**, **testing**, and **repetition**. Spontaneous generation was replaced by biogenesis, based on science. Life can only come from life. How did he do it? By actual science – observation, testing and repetition of the various tests he proposed.

So, when it comes to the age of the earth, are we dealing with **Science? History? Philosophy? Consensus?** Keeping these four ideas straight will help us work through the tangled web of ideas and facts when it comes to dealing with issues like the age of the earth and the history of man. And we must learn to ask, "How do you know that? What of that statement is supported by science?"

Learn these four powerful, yet basic words and how to apply them. Practice by picking up a book or magazine article and dissecting it. Look for statements that teach **science**, **history**, **philosophy**, and **consensus**. In order to keep science pure, we must learn these valuable words.

Questions:

1. Briefly describe the main conflict between science and the Bible.

2. Define the words science, history, philosophy and consensus.

3. Why is it important in any discussion about origins to know the difference between science, history, philosophy, and consensus?

4. Can you find statements in the New Testament that indicate Jesus' thoughts on the historical nature of Scripture? Why would this matter?

Four Practical Exercises in Learning the Four Powerful Words
Exercise 1 – The Genealogies and Chronologies of Genesis: Are They Accurate and Reliable?

One of the biggest stumbling blocks for Christians is the age of the earth question. Does the Bible indicate just how old the earth is, and can I trust it? Can the Bible even resolve this issue? I believe the answer is right in front of us, if we will just take the time to explore the Scriptures that address it – in the genealogies and chronologies of Genesis.

Many say that Genesis cannot be used as a chronology to date the age of the earth. To this statement I say, "Why not?" And the answers are varied, but go something like, "Science has demonstrated through radiometric dating that the earth is 4.6 billion years old." Or, "Genesis has gaps in the genealogical record such that we don't know how many years are involved. Therefore, although the genealogies are probably accurate, the chronologies are not." How can I answer this?

Genesis as a genealogy
Let's go directly to the first genealogy in Genesis 5 and camp there. But first, take some time to read Genesis 5 through and write down every place you see a gap in the genealogy.

History, chronologies, and genealogies
After reading Genesis 5, you might want to take some time to make your own chart of the chronologies and genealogies found there. The following charts are ones that I created that are compiled directly from the Scriptures and are meant

to help you understand the historical nature of the Biblical genealogies and chronologies.

Note: Some people have a problem with the long age spans given for the Patriarchs. Remember that these lived before the Flood and genetically were very close to the physically perfect human being, Adam. Given that, it is not unreasonable to think that they truly lived this long. Conversely, if the ages do not mean what they say, then what are we to make of the ages of the Patriarchs when they had their sons?

The Genealogies and Chronologies of the Patriarchs
First Chart

Creation – year 0

The Flood – about 1600 years after the creation

Adam	The first man; age at death = 930 years
Seth	A son of Adam; age at death = 912 years
Enosh	A son of Seth; age at death = 905 years
Kenan	A son of Enosh; age at death = 910 years
Mahalalel	A son of Kenan; age at death = 895 years
Jared	A son of Mahalalel; age at death = 962 years
Enoch	A son of Jared; age when God took him = 365 years
Methuselah	A son of Enoch; age at death = 969 years. Died the year of the Flood.
Lamech	A son of Methuselah; age at death = 777 years. Died a few years before the Flood.
Noah	A son of Lamech; age at death = 950 years, Lived for 344 years after the Flood

The Historical record from Adam to Noah

The Genealogies and Chronologies of the Patriarchs
Second Chart

Patriarch	Age when he had the patriarchal son	Patriarchal son	Age of Patriarch at death
Adam	130	Seth	930
Seth	105	Enosh	912
Enosh	90	Kenan	905
Kenan	70	Mahalalel	910
Mahalalel	65	Jared	895
Jared	162	Enoch	962
Enoch	65	Methuselah	365
Methuselah	187	Lamech	969
Lamech	182	Noah	777
Noah	Not mentioned	Shem, Ham, & Japheth	950

The Historical Record from Adam to Noah

Genesis 5:1 states, "This is the book of the generations of Adam...." The word *book* is from the Hebrew word meaning *document* or *scroll*. Note the words: "This is *the* book of...." This is a definitive statement. It is authoritative. So, what we are about to look at is not to be taken lightly or as an unimportant writing of some kind. And we do not have the freedom to change any of its meaning.

The next word of significance is the word, *generations*. The word means *a history, a lineage, a family register, or record*. It is the genealogy of Adam through Noah. Just how important is this family register? Is it just an item of curiosity? This is not just some passing list of persons in this record. This is the critical record of Adam's lineage from himself as the very first man, made in the image of God, created out of nothing, to Noah's sons through Seth, one of Adam's sons. After the

Flood, another record picks up the lineage of Noah's sons starting with Shem and finishing with Abraham in Genesis 11.

There is something else that is critical to see here and that is that this genealogy comes through Seth, not through Cain or any of the other many descendants of Adam and Eve. There is a very good reason for this.

What's in a name?
It is important to note that in the Bible, people were named in response to events that were important or hopes that were desired. When we look at Seth, therefore, we need to notice the meaning of his name. The name Seth means *substitute*. Why is this important? Remember that the very close relationship between Adam and God had been lost in the garden of Eden. In my opinion, I think that Eve, as the one who influenced Adam to disobey, must have been very emotionally affected by her wrong decision. Put yourself in her position. The weight of guilt for what she had brought on the creation of God must have been overwhelming. She said of Seth, "God has appointed me another offspring *in place* of Abel, for Cain killed him." (Author's emphasis.) Why, of the many sons and daughters that Adam and Eve produced, were Seth and Abel mentioned? What was so important about them?

Throughout Scripture we are told that Abel was a righteous man and that his sacrifice pleased God. This is the same idea expressed throughout the book of Leviticus when the sacrifices were offered to God in His way according to His word – the offering rose as a *sweet savor to the Lord*. In light of sin and its disastrous consequences, the sacrifice pointed to a time when the Anointed One would come, who would be *the Lamb of God who would take away the sin of the world*, John 1:29. But Cain persecuted Abel for his faith and his sacrifice to God, and he killed him. Remember what Eve said: She said of Seth, "God has appointed me another offspring in place of Abel, for Cain killed him."

In Genesis 4:26 we are told that it was during Seth's lifetime that men began to call upon the name of the Lord. Could it be that Seth represented another attempt to produce a lineage that would be known for its faith in the true God? No matter the reason, without the specific mention of Seth and his lineage, we would not have a Messiah. This genealogy is important.

Genesis as a chronology
The other thing to notice about this genealogy in Genesis 5, is that it is also a chronology. In other words, it not only establishes relationship between Adam and his sons, it also establishes the exact year in which subsequent children and grandchildren were born. This is quite different than the genealogies in the Gospels. There we see relationship only. But here in Genesis 5, we see the all-important element of time involved in producing the genealogy of Adam down to Noah's sons.

Notice that in Genesis 5 the age of the father is given when he had the son of the lineage in question. So, Adam was 130 years old when he begat Seth. Seth was 105 years old when he begat Enosh, and so on. In addition, the total number of years the patriarchal father lived after he *begat* his son and the total number of years the patriarchal father lived is given. This is both a tight genealogical and chronological chain meant to record a very important line of descent from Adam to Noah's sons.

The record in Genesis 5 is a descendant's particular lineage and it is treated very carefully and importantly. There cannot be anything missing. My guess is that Adam and Eve wanted to make sure the truth of what had happened in the garden should never be forgotten. And this would happen through Seth.

Why is this chronology important?
I believe there are two obvious answers. First, it was important to keep the truth alive that Adam was the first man created by God in God's image and likeness.

The second reason is found in one of the names of Jesus, the *Son of Man*, used throughout the gospels. Jesus was uniquely the Son of Man because he was the promised seed of Eve who would crush the serpent. For Jesus to be this promised one, he would have had to show his lineage to the first man and woman, Adam and Eve. Jesus, like Adam, the man, is the image of the one true God (Colossians 1:15) and Jesus, like Seth the man, was the substitute promised to Eve. In fact, Romans 5:14 speaks of Adam as the *type of Him who was to come* (Jesus, Romans 5:14). The Bible also speaks of Jesus as the substitute for our sins (Hebrews 2:17). If there are any missing relatives, how can we confidently establish this truth that Adam was *the type of Him who was to come* (Jesus)?

Genealogy and Chronology
There are four critical things we can say about this genealogy and chronology in Genesis 5:

1. The genealogy is of the first man, for it says that this genealogy starts with Adam in the day that God created man. Human genealogy cannot go back any further than this.
2. For a lineage to have any meaning and importance, there must not be any links missing. Everyone understands this, especially when making claims to royalty. That should, in and of itself, take care of any doubts about gaps in the Genesis 5 genealogy. For it to be an important lineage, it must be complete. If there is a father or son missing, then the genealogy really has no meaning and is worthless.
3. In Jude 14 the Bible tells us in no uncertain terms that Enoch was the 7th generation from Adam. So, we know for sure that there were absolutely no gaps in the first seven generations of mankind through Enoch. There are four generations remaining after that– Methuselah, Lamech, Noah and Shem. If there are going to be any gaps, they could only be among these last four and not in the first.
4. According to these records, Adam lived to see Lamech born. For this to have taken place, there cannot be any gaps in the record, for it tells us that Adam lived a total of 930 years. There are a total of 11 generations from Adam to Noah's sons. If Adam lived to see Lamech, as the Scripture states, then he lived for 9 of the 11 generations recorded. So, that makes the first 9 generations iron-clad with no breaks or gaps!

Crunching the numbers

Let's do a simple math exercise. If we add up the number of years from each patriarchal father to patriarchal son, all the way down to Noah's sons, we get the figure of 1,556 years from the birth of Seth to the birth of Noah's sons, about one hundred years before the Flood. There is no room for a gap, as the age of the patriarch is given at each birth of the son of the lineage. It is meant to be an unbroken lineage with specific years and specific information that would ultimately be used to establish the lineage of the promised seed who would bring ultimate rest and peace on Earth. (It is interesting to note here that Noah's name means rest, or peace. Noah was also a picture or type of that One who would bring peace on Earth.)

We can also do some calculating to see if we can set a Biblical framework for the age of the earth. If Adam was the first man created by God, and Adam was created on the sixth day of creation, then it is clear that the earth was about 1,556 years old at the birth of Noah's sons. The remainder of the genealogies given for Shem to Abraham to Jacob to Jesus are recorded in the Bible. If we assume the accepted date of the birth of Jesus somewhere between 6 and 4 BC, we can add approximately 2000 years up to the present to get a little over 6,000 years for the age of the earth. It is really quite simple. (By the way, the challenge in fixing the birth of Jesus is not a Biblical issue. It is a problem with fixing the exact rule of Herod, the king, and when he died. The promise of the 1st coming of the Messiah was fulfilled at the birth of Jesus.)

So, if we are to trust this unbroken lineage, then the time that man has been on the earth would be closer to 6,000 years. But scientists tell us that Homo sapiens, meaning *wise man*, came on to the scene around 200,000 years ago. If this is correct, then we have a lot of gaps to fill in our Biblical chronology. Effectively, if we are to find a way to harmonize the Scriptures and the secular viewpoint, we have to find some way to account for about 194,000 years of missing genealogy and chronology.

Let's do a little more math to see if we can fix this discrepancy. We know that there are certain dates in history that we can set without any doubt because of the records that we have. We are, therefore, primarily concerned with dates associated with those first 11 generations, up to Shem. For us to get to 200,000 years, to harmonize the Scriptures with this amount of time, we need to create some gaps in these early genealogies and chronologies. If the average man lived for 800 years (using our record in Genesis as a guide), then there would have to be at least 200 father/son connections missing in the Biblical genealogy/chronology of Adam to Shem! It is not just one or two possible missing persons. It is at least 200 connections. If that is the case, if we are to trust the scientific idea that man has been around for 200,000 years, then who should even pay attention at all to these worthless genealogies in Genesis 5, let alone the rest of the Bible? What good does that kind of genealogy and chronology serve?

The main challenge since the 1800s to this extremely important record, has been a contradictory voice coming from those who claim to be true scientists, who say that the earth is 4.6 billion years old. Most people uncritically accept this age for the earth as fact. Remember from our second lesson, however, that this age for the earth is really a matter of philosophy and not scientific facts. For the believer, faith in the revealed genealogies and chronologies is critical. Without faith, we cannot please God (Hebrews 11:6). Abraham demonstrated true faith as recorded in the Bible: "Abraham BELIEVED God and it was credited to him as righteousness." (Romans 4:3, author's emphasis.)

Righteousness was credited for believing what God had spoken, not just for believing that a god exists. But if we accept the age of the earth as insisted on by scientists, then the Biblical faith, the kind the Bible says pleases God, is destroyed. Practically speaking, our day-to-day faith in God and His Bible will be stifled again and again and soon other parts of the Scripture will become suspect. If Genesis is wrong about its chronologies and genealogies connecting Adam to Shem, then could the Bible be wrong in its presentation about other important historical events and records? This is vitally important because ultimately this is tied into the physical resurrection of Jesus Christ.

Questions:

1. Construct your own chart that records the lineage from Adam to Noah's sons. Add up the years that connect each patriarch and his son. Is there any way that more descendants can be added?

2. If the chronologies are in error in Genesis 5, list at least 4 consequences of this error.

3. Are these consequences important? Why or why not?

4. When studying the genealogies and chronologies of Genesis, how would you apply the four powerful words in your study?

Four Practical Exercises in Learning the Four Powerful Words
Exercise 2 – Time and Chronology in the Secular Geological Age

The subject of time and chronology is important to the secular geologist. It should be to the Christian too. In fact, the entire secular geological system is based on time and a secular chronology of Earth history. And modern geology guards this like one would guard a precious jewel. Anything or anyone that would challenge the current secular concept and teaching of vast ages of time will be criticized and ostracized from its circles of discussion.

James Hutton, and geochronology
The subject of time in modern geology is so important that a separate branch of geology is devoted to it. The study of geological time in modern geology is called *geochronology*. Geochronology really had its beginnings in the late 1700s when James Hutton, the Father of Modern Geology, often studied the weathered rock layers in Scotland.

Hutton's framework of thinking was Deism. Deism was a product of the Enlightenment. Hutton was one of the Fathers of the Scottish Enlightenment. The Enlightenment insisted on rejecting things like holy books and church doctrine. Although they generally believed that a god had created all things, Deists believed that God made man to think and reason for himself. This meant that in looking at the rock layers, the historical account of the Genesis Creation and Flood was not to be automatically accepted, but to be judged by man's thinking and/or to be jettisoned as myth.

Hutton, therefore, imagined that it took a long time for the rocks to accumulate and harden as sediments – much longer than what the Bible had stated. This idea found great appeal among the intellectuals of the time because it did not include anything to do with the Bible. And thus, the idea of an old Earth found its way into the modern view of Earth science as *geochronology* – the study of the great ages of Earth's past history. Is this science, history, philosophy, or consensus? It has very little to do with science and history. Geochronology is mostly philosophy and consensus.

What has amazed me in this whole debate is the way that most of the modern church handles the discussion of time as it relates to Earth history. Most of the modern church believes that science has demonstrated beyond a reasonable doubt that the earth came about billions of years ago. Most of the church treats this subject with little interest, saying that it is really not all that important. But the secular geologist thinks it is a big deal and is a firm believer in the idea of an old Earth. He takes great pains to interpret everything he sees in light of this old Earth framework. Accordingly, modern geologists know their chronology much better than we Christians know ours. Modern geologists are dedicated to their chronology and know it backwards and forwards. Let's take a look at this chronology.

The Geologic Column or Time Column

During the early to mid-1800s, rock layers began to be studied by individual naturalists and assigned names according to the particular geographical area in which they were studied – some in England, some in France, some in Russia, and some in Germany. It was quickly noticed that these rock layers contained fossils. But these layers of rocks were not stacked on top of one another but were quite independent and isolated from one another. The big question became what to do with these different layers of rock and fossils. Are they related in some way? How should they be organized? In other words, they were not *correlated*. The following are the names for these different areas and where that name came from[1]:

> **Cambrian** – From a classical name for Wales
> **Carboniferous** – From the coal beds in England, name based on study of British rock formations
> **Jurassic** – After the limestone in the Jura Mountains
> **Cretaceous** – From Latin, creta, meaning chalk, of the Paris Basin (France)
> **Triassic** – Red beds, capped by chalk, followed by black shales
> **Permian** – After rocks in Perm, Russia
> **Silurian and Ordovician** – After ancient Welsh tribes
> **Devonian** – After the English county of Devonshire

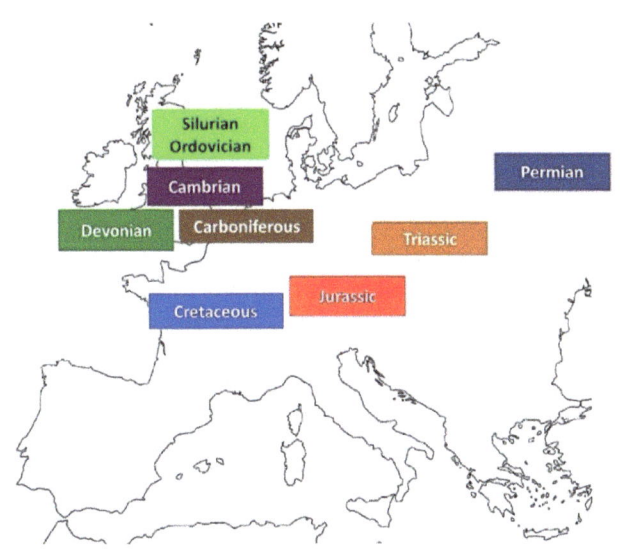

The Geological Picture of Europe in the 1800s

The concept of correlation
Correlation is a mutual relationship or connection between two or more things. So correlation and the organization of the different rock layers became the goal of the 1800s and indeed is still the goal today as more and more fossils are discovered that do not fit the standard picture that was developed before the turn of the 19th Century. It also must be pointed out that radiometric dating did not exist until the 20th Century. So, how did early geologists assign ages or dates to the fossils? There was no objective scientific way to date the rocks or the fossils!

By the mid- to late- 1800s and especially after Darwin published his book on origins and development of living things in 1859, the individual rock layers were organized around one common principle – simple creatures called protozoa, meaning, *first life*, must have changed over time; new, more complex creatures, called *metazoa*, multi-cellular life, must have developed as a result. Remember that DNA and genes were not known in these days. If a plant or animal looked simple, then it was simple. The rock layers containing these simple creatures as fossils along with other fossils began to be organized into a hypothetical column that became known as The Geologic Column. The "simple" creatures, like protozoa (one-celled creatures) belonged on the bottom of the column. And more "advanced" creatures, like dinosaurs, belonged further above them. This is, of course, a belief and not actual science. Without the organizing principle of evolution (genealogy) and time (chronology), The Geologic Column would not exist! And without the belief in evolution, rock layers are simply rocks with fossils in them.

Cenozoic Era	←	Age of mammals
Cretaceous	←	Pterosaurs, dinosaurs dominant
Jurassic	←	First birds, theropods
Triassic	←	First dinosaurs
Permian	←	Rise of reptiles; deserts
Carboniferous	←	Amphibians; land plants flourish
Devonian	←	The Age of Fish; first land plants
Silurian / Ordovician	←	First bony fish
Cambrian	←	First vertebrates; simple life on ocean bottom

Goal of the 1800s - Correlation

Correlation was based on the IDEA that life developed from simple to complex over long periods of time. There was no scientific way to date the rocks or fossils.

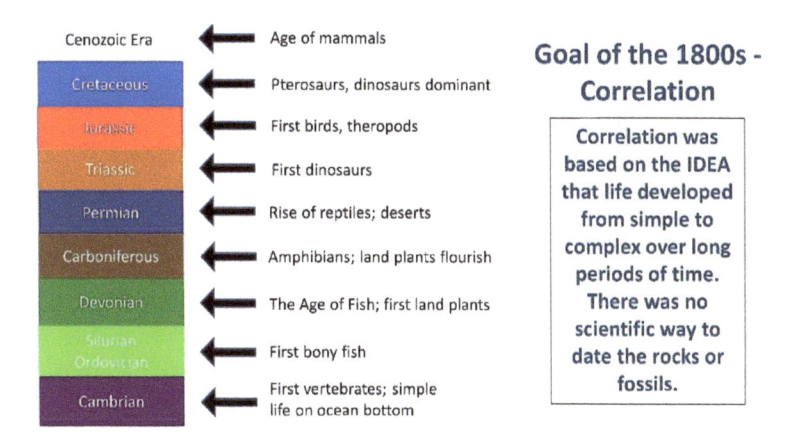

This new concept of Earth history became thoroughly established in the scientific community by the late 1800s, not because of science, but because the only other alternative viewpoint was that of the Bible. And the Enlightenment had taught us that no church, Bible, or religious dogma would influence how scientists would interpret the rocks and fossils. This stance obviously closed off any further discussion and has produced a sort of straitjacket perspective on Earth history within the geological community. Geologists have continued to miss the catastrophic global flood of Genesis as possibly a better explanation of many geological landforms. But that is not a discussion that any modern scientist wishes to even entertain. In fact, there are some scientists who consider such viewpoints to be criminal if taught to children.

Fossils and the Geologic Time Column

Early geologists noticed that certain rocks contained fossils while others did not, even though they were the same rock type. Rocks with fossils of what were considered at that time to be strange marine creatures were first recognized by Adam Sedgwick in 1835 in Wales. He gave these rocks the name of Cambria or Cambrian, after Cambria, the Latinized form of *Kumree*, the Welsh name for Wales.

Geologists also noticed that below these particular rocks were rocks that did not contain any fossils, even though many of the rocks were sedimentary. These geologists found that the Cambrian rocks contained strange marine fossils such as trilobites (extinct), corals, brachiopods, and crinoids (some of which are extinct). They also found that there were no fossils found in the rocks resting directly below the Cambrian. They reasoned, therefore, that the Cambrian must be the oldest of the fossil-bearing rocks. So, all rocks containing these types of creatures were assigned to the Cambrian system of rocks and were considered to be ancient.

The rocks below the Cambrian were initially called Pre-Cambrian and are still referred to by that name today. Later, these rocks also became known as Proterozoic rocks, meaning, *earlier life*. The addition of this geologic era to the Geologic Column was necessary because evolutionary thinking demanded that there be evolving life before the Cambrian. In other words, the ancestors to the Cambrian creatures must be located somewhere in these lower rocks. In fact, the Cambrian became a big headache for Darwin. To this day the Cambrian continues to be noted for the sudden appearance of complex invertebrate marine life – the most prolific and abundant of any fossil bearing strata. And further collecting has demonstrated that representatives of every phylum of life has now been found in the Cambrian Rocks.

This mystery concerning the prolific number of fossils in these rocks has been called the *Cambrian Explosion*. It remains one of the greatest mysteries in modern geology. No indisputable fossil transitional form has been discovered in the Precambrian.

The naming of other geologic systems followed and gradually a well-formed alternative picture of the history of life was developed, ultimately leading to full-fledged evolution by 1859.

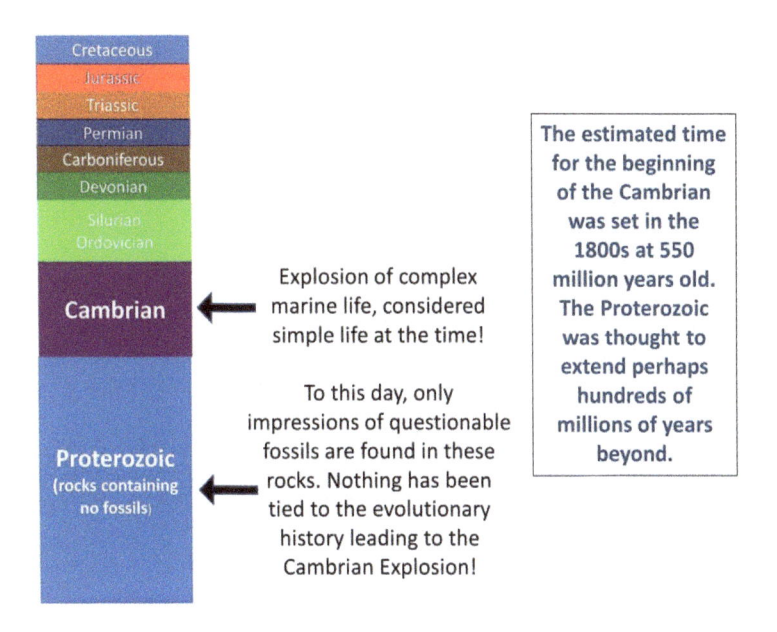

Cretaceous	
Jurassic	
Triassic	
Permian	
Carboniferous	
Devonian	
Silurian	
Ordovician	
Cambrian	← Explosion of complex marine life, considered simple life at the time!
Proterozoic (rocks containing no fossils)	← To this day, only impressions of questionable fossils are found in these rocks. Nothing has been tied to the evolutionary history leading to the Cambrian Explosion!

The estimated time for the beginning of the Cambrian was set in the 1800s at 550 million years old. The Proterozoic was thought to extend perhaps hundreds of millions of years beyond.

The Geologic Column became increasingly the Geologic Time Column, attempting to explain all of Earth history, not just early life on Earth. Today the Geologic Time Column is treated as a sort of the "Bible" of modern geology. Even radiometric dates are adjusted, accepted, or rejected based on this column that was developed in the 1800s. The Column has been added to since those days and it reads like the Bible does, with a fairly complete genealogy and chronology, presumed factual proof for the origin of life and the geology of the ancient Earth.

The modern column usually begins at 4.6 billion years ago, and has been presented in many different ways. The following representation[2] begins in the lower right at approximately 4.5 billion years ago. It is clear that this table represents a replacement, intended or not, of the Biblical Chronology.

EON	ERA	PERIOD	EPOCH		Ma
Phanerozoic	Cenozoic	Quaternary	Holocene		0.011
			Pleistocene	Late	0.8
				Early	2.4
		Tertiary (Neogene)	Pliocene	Late	3.6
				Early	5.3
			Miocene	Late	11.2
				Middle	16.4
				Early	23.0
		Tertiary (Paleogene)	Oligocene	Late	28.5
				Early	34.0
			Eocene	Late	41.3
				Middle	49.0
				Early	55.8
			Paleocene	Late	61.0
				Early	65.5
	Mesozoic	Cretaceous	Late		99.6
			Early		145
		Jurassic	Late		161
			Middle		176
			Early		200
		Triassic	Late		228
			Middle		245
			Early		251
	Paleozoic	Permian	Late		260
			Middle		271
			Early		299
		Pennsylvanian	Late		306
			Middle		311
			Early		318
		Mississippian	Late		326
			Middle		345
			Early		359
		Devonian	Late		385
			Middle		397
			Early		416
		Silurian	Late		419
			Early		423
		Ordovician	Late		428
			Middle		444
			Early		488
		Cambrian	Late		501
			Middle		513
			Early		542
Precambrian	Proterozoic	Late	Neoproterozoic (Z)		1000
		Middle	Mesoproterozoic (Y)		1600
		Early	Paleoproterozoic (X)		2500
	Archean	Late			3200
		Early			4000

If I am going to be a consistent and serious student of the Bible, I must become convinced that the Bible is God's revelation to man of his genealogical and chronological history and that it is accurate in its every detail. I must become as familiar with Biblical genealogy and chronology as the secular geologist is with his. I must discover where the modern representation of Earth history differs from that of the Bible and formulate both a Biblical historical framework and an ability to use it to interpret the rocks and fossils.

Today the Biblical chronology with its young portrayal of Earth history is considered by many scientists, lay people, and even many Christians to be a joke. Many believers are comfortable with a sort of dual belief system – parts of the Bible are true, and parts are not. I believe this thinking has saturated the church basically for two reasons: (1) We have been so bombarded with the modern geologic history of Earth that it seems normal to us. (2) We don't know our own history. We don't know it well enough to counter the modern view or to even present a defense for our own story. To be an apologist for the Christian Faith, we must become better equipped to do both. Our entire belief system rises or falls on an accurate Earth history as portrayed in the Biblical genealogies and chronologies.

Rebuilding a Biblical framework for interpreting geology

Is it even possible to construct a time column based on the Bible? If it is history, it can and must be. A reliable historical account based on documents and eyewitnesses is the only way to build a dependable framework in which we can interpret what we see around us in the rock formations and landforms.

A Basic Biblical Framework

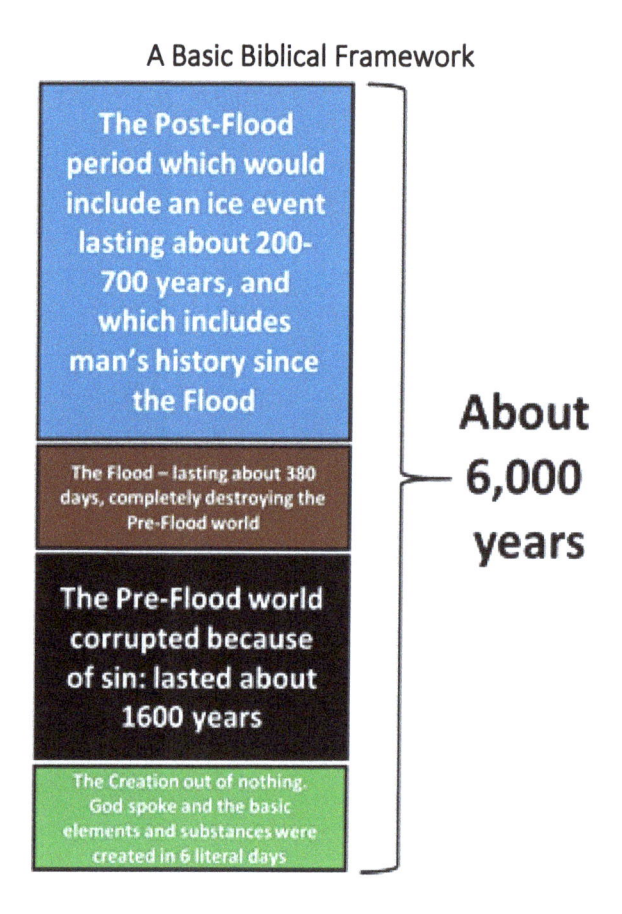

The Post-Flood period which would include an ice event lasting about 200-700 years, and which includes man's history since the Flood

The Flood – lasting about 380 days, completely destroying the Pre-Flood world

The Pre-Flood world corrupted because of sin: lasted about 1600 years

The Creation out of nothing. God spoke and the basic elements and substances were created in 6 literal days

About 6,000 years

By taking a straightforward understanding of the book of Genesis and adding up the years given for the chronologies of the Patriarchs, one can establish a basic historical framework for interpreting Earth history. It has always been amazing to me just how effective this simple tool has been at putting the pieces of geology together into a coherent whole. Even so-called mysteries of modern geology, like planation surfaces and unconformities, can easily be explained by following this simple Biblical framework.

Taking a straightforward reading of the Book of Genesis, a completely different view of Earth history comes to the forefront. What a drastic difference between the philosophy of modern geology and the history of the Book of Genesis!

Questions for discussion:

1. What is the field of study that concerns time in modern geology?

2. Before radiometric dating, what drove the development of long ages for Earth history?

3. Why is the Cambrian system of rocks called *the Cambrian Explosion*?

4. What is the difference between the terms *Precambrian* and *Proterozoic*?

5. Apply the four powerful words to the study of chronology in both the secular and Biblical understanding.

Four Practical Exercises in Learning the Four Powerful Words
Exercise 3 – Evolutionary Gaps in the Fossil Record: How Serious Are They?

Why are the fossils that are buried in the layers of rock all over the earth such an important issue in modern geology? Remember that we said that much of modern geology falls outside of science. In other words, most of the 4.6 billion years of Earth history has not been observed nor recorded by human observation. Scientists use their power of interpretation to tell a story based upon what they are observing now. That doctrine was developed in the late 1700s and is called *uniformitarianism*. It is the foundation of all of modern geology. Since there are no written records that tell us exactly what happened in the distant past, geologists must rely on the rocks, fossils, and their interpretive framework of uniformitarianism to reconstruct a history of the earth. They do this by placing their template or framework of uniformitarianism over the rocks and fossils. But just how much of uniformitarianism is actual science? The science is only that which can be observed, tested, and repeated!

When we talk about fossils, we are in the field of paleontology, which is part of the study of geology. When you study paleontology, it is a given that you accept the uniformitarian worldview. Paleontology accepts the dating of the rock layers, based on its uniformitarian worldview. The dates for the layers of rock come from the fossils that are found in them. And fossils are dated according to the age of the layers of rock. The fossils then become the proof that the layers of the rocks are ancient; this is classical circular thinking!

This would not be a problem if there was some other record that was used to confirm their conclusions, like a written record of eyewitness accounts. But as it is, the only record we have is what we can observe in the present – petrified remains of plants and animals buried in the rocks. There are no dates printed on the rocks, no eyewitnesses, and no written records. There is no way to test the past, and no way to repeat what happened.

Just what do geologists do with fossils? They approach them, in a sense, by faith. They believe that the idea of uniformitarianism is sufficient to tell them what happened to explain the origin of fossils. Geologists have faith in the following:

- They believe that the rocks are ancient – in 1999 the oldest rock on Earth was supposedly dated at 4.03 billion years and the oldest fossil has supposedly been dated at around 3 billion years old
- Evolution of life from simple to complex is believed to be a scientific fact as shown by the fossils
- They believe that all of Earth's geological processes have been and are naturalistic processes and neither miracles nor a belief in God have any place in science

Harmonizing the Scriptures and evolutionary ideas
There are those that do try to harmonize the Scriptures and science. BioLogos is one group that seeks to do this. On their web site, it states: "Fossils open a window deep into the history of the earth. Through that window we learn about how whales evolved from four-legged creatures to the

aquatic animals we know today, we learn about our own species and where we came from, and we learn more about God who made it all."[1] Since geologists are adamant that fossils show the evolutionary progression from simple to complex, it follows that geologists **must be able** to show this in the fossils.

The following is a list of generally accepted fossil evidence among evolutionists. This evidence is used as confirmation that evolution is a fact. (Please note, the ages given are their interpretations, not actual observations.)

- *Fossils of single-celled organisms have been found in rocks greater than 1 billion years old*
- *Fossils of multicellular organisms have been found in rocks around 550 million years old*
- *Fossils of fish without jawbones have been found in rocks around 500 million years old*
- *Fossils of amphibians have been found in rocks around 350 million years old*
- *Fossils of reptiles have been found in rocks around 300 million years old*
- *Fossils of mammals have been found in rocks around 230 million years old*
- *Fossils of birds have been found in rocks around 150 million years old*

This list shows an evolutionary progression from simple to complex and in chronological order. In effect, there is both a genealogy and chronology, just like the Bible's, only much longer. And this evolutionary progression is taught in the high school and college classroom and has been since the days of Darwin in the mid-1800s. If this account is correct, then, of course, everything that the Bible teaches is not only incorrect and unscientific, it is a downright lie. So, we can't just ignore it.

What is needed, however, is to show conclusively that the rocks are indeed the ages ascribed to them by geologists and that these rocks contain a progressive evolutionary descent in the fossils. It is precisely here that modern geology has had a checkered past. The Bible has a written historical account stated in the book of Genesis concerning the Creation and the Flood. Secular geology has no written history of the evolution and development of the earth. Secular geologists rely on fossils that they think form the record for evolutionary age and progression of life.

Virtually all geologists today tell us that it is radiometric dating that has determined the ages of the rocks and that that is the science. But radiometric dating did not exist until the 20th Century. The great ages for the earth's processes, rocks, and fossils were worked out in the 1800s. If there were no eyewitness accounts or historical records that verified these ages, then what did early geologists rely on? It certainly wasn't radiometric dating. It was a philosophical belief system, called the Enlightenment, that shaped modern geology. Radiometric dating came much later, almost a hundred years later, and even it has to be selectively handled to coordinate with what geologists had already concluded in the 1800s.

What do the fossils show?
Now, about those fossils. What do they really show? The list below is a general description of what we actually observe in the fossils.

1. The rock layers that contain virtually all of the fossils are sedimentary in type. Sedimentary rocks were predominantly laid down by water and mud. This is a global phenomenon. 70% of the earth's surface is covered with sedimentary rocks.
2. The typical rock layer that contains fossils, mostly shows mass death and burial. Further, most of the rock layers that contain dinosaur bones show horrendous disarticulation by way of some catastrophic event or events. Disarticulation is the tearing up and spreading of bone material. Disarticulation is characteristic of the fossil bones we find all over the globe. When I studied geology years ago, the doctrine of uniformitarianism was applied in a very strict gradualist way. It would not permit the discussion of catastrophism. Today, more and more geologists are appealing to local catastrophic events to explain the rocks and disarticulated fossils we see. But they reject the idea of a global catastrophe.

3. As these rock layers with fossils tell of catastrophic events that shaped them, they also tell us of extinction on a grand scale. In fact, the fossil record is really a record of death and extinction, not a record of evolution.

4. Some fossils of plants and animals we thought had totally disappeared millions of years ago suddenly reappear without any change in subsequent layers of rocks, totally blurring the lines of any sort of progressive ancient geologic time between them.

5. Scientists often refer to index fossils as reliable time markers in the geological record. These index fossils are fossils that first appear in certain rock layers, and then disappear in subsequent layers. Therefore, if fossils are found with these particular index fossils, it is concluded that those fossils belong to that particular age. The problem comes in when so-called index fossils have subsequently been found in rock layers below or above the designated index fossil. This throws off the whole evolutionary tree. Because index fossils are used as time markers for rock layers, then finding them in what geologists call, *older rock layers*, or *younger rock layers*, changes the entire evolutionary story. These misplaced index fossils do exist, and they are not rare.[2]

6. Geologists have sorted the fossils into a complex order and arranged them into a device called, the Geologic Time Scale or Column (*Exercise 2*). This device does not appear in its entirety anywhere on the earth. But geologists insist that this device, developed over a period of time in the 1800s, accurately portrays the evolutionary parade of life. This arrangement is purely because of a uniformitarian, evolutionary interpretation of the rock layers and fossils.[3]

7. Darwin thought that life had progressed over millions of years from simple to complex. Geologists advanced this idea, believing that the fossils would eventually be found, justifying Darwin's ideas. What have geologists found in the fossils to date? Present observation of living things and petrified life in the fossils show that all of life, in its various forms, is

extremely complex in its structure. Darwin did not know about DNA. Today, no life is considered simple. It is all complex. To date, no conclusive fossil ancestors to the complex forms found in the fossils have ever been found.

Although some secular scientists tell us that this phenomenon is because of the incomplete nature of the fossil record, that is a matter of interpretation. At the same time, other scientists acknowledge the marvelous preservation and abundance of the fossil material in the record.

If evolution had been proven to be a fact, then yes, the fossil record connecting the various species would be considered very poor. But that is putting the cart before the horse. There are no connecting fossil links demonstrating the evolution of life over hundreds of millions of years! This has baffled many scientists because it seems that only those species that have evolved and survived have been preserved. What happened to the many billions of transitional fossils that must have lived but are nowhere to be found. Why did they not survive fossilization to finish the story? Not one indisputable fossil of a transitional form has ever been found.

A written, historical account
In light of the historical Genesis record of a special creation, it would make most sense to interpret the abundance of well-preserved fossil plants and animals as having been created by God who made them with the ability to reproduce variations within genetic boundaries – *kinds*, the Bible calls them.

And in light of the historical written record of a global Flood, it would make most sense to interpret the fossils of mass burial and disarticulation of plants and animals in the rock layers all over the earth, as the result of a global Flood which would have been a violent and destructive event.

Although fossils are not a written record, the fossils do serve as a witness to the written history of Genesis – that God created special living things and then subsequently destroyed most of them in a global Flood. The only land-dwelling, air-breathing creatures to survive were those who were kept on Noah's ark during the global and catastrophic Flood. The state of the fossil evidence we have today does verify the Genesis written record of history.

The abundance of fossil kinds with variation and the abundant lack of transitional fossils is a serious problem for modern geology. There should be an abundance of transitional fossils if there is an abundance of the finished product. After all, evolutionists teach us that the process of evolution is a chance process. Chance processes would require the abundance of trial and error creatures that did not make the cut. Therefore, we should expect to find an enormous number of transitional fossils that fill out and, in some cases, complete the evolutionary story. But there are simply none of these! Millions of fossils of distinct kinds of plants and animals exist with variations. But no indisputable fossils have been found establishing the science of transitional forms. When it comes to evolution, are we looking at proven evolution with plenty of evidence, or are we looking at an idea desperately being clung to with the hopes of finding the "missing" links?

Questions:

1. The doctrine that the earth has developed naturally over billions of years and according to observed geological processes is called what?

2. In what ways is this doctrine science? In what ways is it not science?

3. Briefly describe the general nature of the fossil/rock record.

4. Why is the lack of transitional fossils between the various kinds a problem?

Four Practical Exercises in Learning the Four Powerful Words
Exercise 4 – Dinosaur-to-Bird Evolution, the Story that Never Seems to Die

In 1861, just two years after Darwin's book, **On the Origin of Species by Means of Natural Selection, or the Preservation of Favored Races in the Struggle for Life** was published, a fossil feather looking exactly like a bird feather was discovered. It was given the name, *Archaeopteryx*, meaning, *ancient wing*. Most of the 11 specimens of Archaeopteryx that have been discovered since, come from the Solnhofen limestone in Bavaria, southern Germany. Geologists call this formation a *lagerstätte*, meaning, *storage place*. These are rather common and remarkable geological formations known for their superbly detailed fossils. These deposits in Germany have been dated at 150 million years old. No other indisputable bird fossils have been found older than Archaeopteryx, according to the evolutionary chronology. In fact, the Germans call Archaeopteryx, *Urvogel* (meaning *original bird* or *first bird*.) Archaeopteryx has held that distinction since 1861, over 150 years!

Just a reminder: I am going to be talking about a lot of secular dates for the various candidates for dino-to-bird evolution. These dates are a product of philosophical interpretation only. They have never been established by science to be true.

Geologists tell us that Archaeopteryx was one of the most important fossils ever discovered. Almost all paleontologists acknowledge that it possessed the power of flight and is today considered to be fully bird. The interesting thing about Archaeopteryx, however, was that it possessed teeth and claws, unlike any bird found at the time of discovery. Thomas Huxley, known as, *Darwin's Bulldog* in the 1800s because of his fanatical support of Darwin, thought that Archaeopteryx must have been a transitional creature between a small theropod dinosaur called *Compsognathus* and modern birds because of these features. This idea stuck and today Archaeopteryx is still considered one of the best examples of a transitional fossil in the gap between dinosaurs and birds. Indeed, it is still considered to be one of the best evidences for a transitional link!

What was Archaeopteryx?
Was Archaeopteryx a transitional creature between dinosaurs and birds? Since Archaeopteryx is considered by paleontologists to be one of the best examples of a transitional fossil, it would be good for the apologist to know about this fossil critter.

Ornithologists, those who study birds, agree that Archaeopteryx was a good flier. But many disagree about his status as a missing link. Was Archaeopteryx simply an example of the extinction of a bird that could fly just like our birds of today, but had a few extra interesting characteristics that we don't find in modern birds? That is a question that science cannot answer, because we don't have Archaeopteryx around today to observe how he behaves. His status as a missing link must then simply be a matter of interpretation. At this point Archaeopteryx can legitimately be classified as an extinct bird with teeth.

The status of Archaeopteryx as one of the oldest and best-established missing links has had some very interesting back stories. One particular publication caught my attention several years ago and deserves our attention.

Several years ago, **National Geographic Magazine** published an artist's rendition of the evolution of the bird's wing. It was entitled *The Path to Birds*.[1]

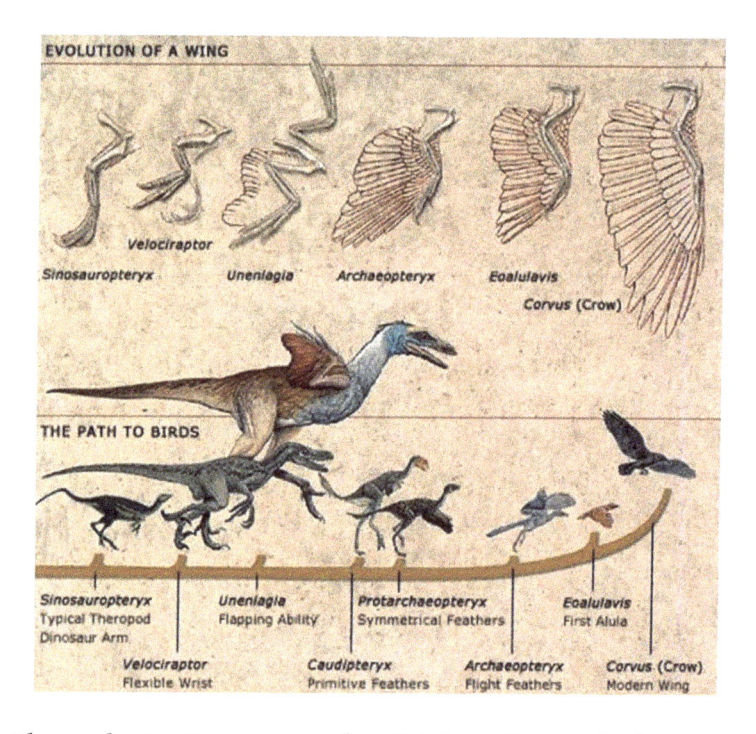

The evolution into our modern bird, ending with the crow, is clearly depicted. The original artwork does not give any dates for the various candidates, just evolution's arrow, indicating how the evolution of the bird developed over time. Here is where the story gets interesting.

Most people will not take the time to research each one of the candidates for the dino-to-bird series portrayed in this

58

picture. A normal person will simply notice the pictures of small theropods evolving into flying birds. The curious question is why doesn't the picture label the *Path* with dates of when these candidates were supposed to have existed? That would be very helpful in seeing fully the chronology of this evolution story. But the author did not do that. So, you and I will have to do it.

Discovering the dates of the various candidates is initially a fun activity. Then it hits you. The dates are all out of order. There does not seem to be a chronological order to the arrangement of the candidates for dino-to-bird evolution. In fact, arranging the candidates according to the actual chronology would indicate that birds evolved into dinosaurs! It's backwards. (Dates are approximate, but within the secularly accepted range for each creature.)

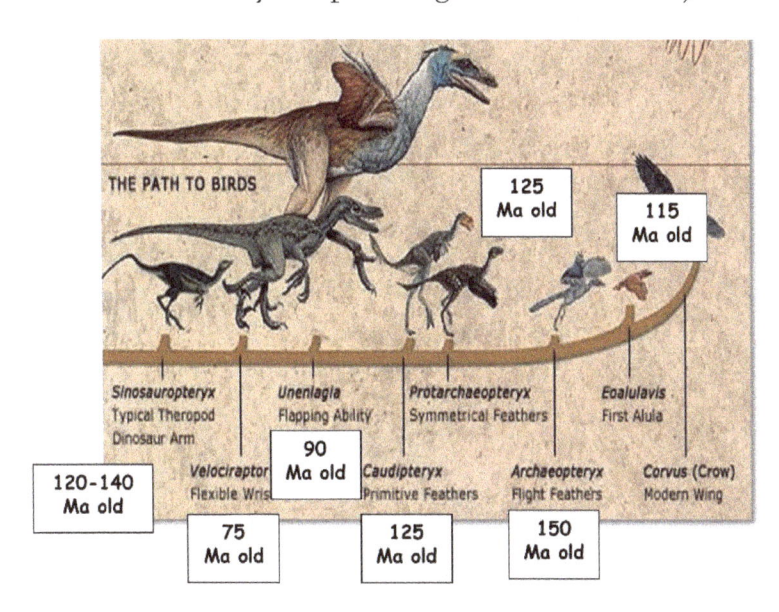

THE PATH TO BIRDS

Sinosauropteryx	Unenlagia	Protarchaeopteryx	Eoalulavis
Typical Theropod Dinosaur Arm	Flapping Ability	Symmetrical Feathers	First Alula
120-140 Ma old	Velociraptor Flexible Wrist	Caudipteryx Primitive Feathers	Archaeopteryx Flight Feathers	Corvus (Crow) Modern Wing
	90 Ma old			
	75 Ma old	125 Ma old	150 Ma old	

125 Ma old

115 Ma old

The celebrated Archaeopteryx is positioned very close to the most recent bird (crow) in this *Path to Birds*. He is placed toward the end of the evolutionary process of dinosaur-to-bird. But he is the oldest of the candidates! Archaeopteryx could fly and was found in rock layers dated 150 million years old. Should he not be placed chronologically at the beginning of this chart? But if we do that, then evolution would seem to be going backward!

Assuming that National Geographic, who paid for an artist's rendition for the *Path to Birds*, knew this same information, you have to ask them, "What were you thinking?" And this highlights the problem with the dinosaur-to-bird evolution story – there are so many transitional parts still missing that the story simply cannot be told, without hiding the dating information. The various theropods and birds are all arranged as if there was a nice neat chronological and genealogical story to this arrangement. But the dates just simply do not tell the same story. The fossils of Archaeopteryx are said to be 150 million years old and indicate that he was fully bird. Whatever else the fossils indicate, they don't agree with the scripted story.

The Bible, birds, and dinosaurs

Aside from the errant chronology, there is the challenge for anyone wishing to know what the Bible might have to say about dino-to-bird evolution. Although the word dinosaur wasn't invented until thousands of years after the Bible was written, the one clear teaching from Genesis 1 in relation to dino-to-bird evolution is that birds were created on day five of Creation week, along with sea creatures. Land-dwelling, air-breathing animals were created on day six of Creation week. So, dinosaurs came *after* birds, not before.

The Hebrew word *behemah* is used for many of the land-dwelling creatures created on day six. This word is the root of *Behemoth*, later used in Job to describe a huge sauropod or long-necked dinosaur. The idea that dinosaurs were created on day six is largely hidden in modern translations of the Bible. The modern translations generally follow the King James Version which translates the word *behemah* as *cattle*. That is simply too narrow and misses the point of the story. The Biblical idea that dinosaurs lived with man, therefore, has been obscured and it becomes an out-of-sight-out-of-mind fairy tale.

Since dinosaurs were considered land animals (in total agreement with secular paleontology, by the way), it is then clear from this passage of Scripture that birds came BEFORE dinosaurs, not as the result of the evolution of dinosaurs into birds. For the Bible believer that should end the matter. Paleontologists have simply misinterpreted the evidence.

In another National Geographic publication, **How Dinosaurs Took Flight** published in 2005, the writers printed a more complete list of the latest evidence for dinosaur-to-bird evolution. Some of the same critters pictured in *Path to Birds* are still featured, but there are some brand-new candidates. We won't go over each candidate in depth, but what I want you to notice in this list is the same mistake that continues to be made: the chronology and genealogy is still out of order and makes absolutely no sense if we are trying to show a simple evolutionary progression from dinosaur to bird. Let's take a look at the list published by National Geographic that is supposed to portray dino-to-bird evolution.

Note: *The creatures listed are the ones included in the book. I have added the years when the particular creature was supposed to have lived. Please note that the ages are not exact, and there is some variation from different resources. But this variation is accepted within the secular scientific community and does not change what this exercise shows: that none of these creatures are older than archaeopteryx.*

In order of appearance and evolutionary sequence, they are[2]:

1. **Sinosauropteryx** (124-122 million years old) the first claimed feathered dinosaur (feathered theropod or dino-bird) – but still younger than Archaeopteryx who was supposed to be the best example of a transitional fossil between dinosaurs and birds.
2. **Dilong** (125-124 million years old) – considered by paleontologists as a feathered Tyrannosaur; again about 25 million years younger than Archaeopteryx. Since Archaeopteryx was already flying and had feathers, Dilong cannot be an evolutionary link between dinosaurs and birds.
3. **Shuvuuia** (75 million years old) – a feathered *dino-bird*; 80 million years younger than Archaeopteryx.
4. **Caudipteryx** (124 million years old) – thought to be a feathered close relative of *oviraptors*, many paleontologists now consider Caudipteryx to be an extinct flightless bird.
5. **Tyrannosaurus** rex (67-66 million years old) – this dinosaur has **not** been found with feathers...yet many paleontologists insist that it most likely had feathers, at least while it was young. Why would he need feathers when Archaeopteryx was using them about 90 million years earlier?

6. **Beipiaosaurus** (124 million years old) – a feathered *therizinosaur*, a type of theropod dinosaur.
7. **Pedopenna** (164 million years old) – a controversial specimen, because the age of the Chinese fossil beds where it was found is hotly debated.
8. **Microraptor** (125-120 million years old) – a four-winged *dino-bird*.
9. **Protarchaeopteryx** (124 million years old) – a feathered *dino-bird* was supposed to come before Archaeopteryx, according to its name, but is 25 million years younger than Archaeopteryx.
10. **Sinornithosaurus** (124-122 million years old) – a feathered *dino-bird*.
11. **Archaeopteryx** (150 million years old) – the oldest and possibly first bird, according to evolutionists, and was completely a bird; he could fly. And notice here that he is again placed at the end of the long evolutionary genealogy, although his chronology is the oldest of the candidates. This is a real head-scratcher!
12. **Eoalulavis** (125 million years old) – listed as an early bird. But even as an early bird, he came before many of his supposed theropod relatives.
13. **Corvus** (crow; 16 million years old) – a living bird.

The various chronologies for dino-to-bird evolution just do not work unless we change all the rules that would make any sense of chronology and genealogy. And that is exactly what evolutionists have done. You and I are probably most familiar with the Linnaean Classification system. You know the one – Kingdom, Phylum, Class, Order, Family and so on. It classifies things according to an objective system. You can observe these characteristics in real life.

But over the last 20-30 years a new system of classification has been emerging called Cladistics. This system is extremely complicated and is hotly debated because the system depends entirely on a subjective view of relational characteristics. For instance, paleontologists are convinced that birds share certain characteristics with dinosaurs and so must be related. So, the candidates are arranged according to these supposed shared characteristics, even though the purported dates of the fossils do not line up.

What kind of scholarship is this? How can you trust this? And that is the state of dino-to-bird evolution – none of the fossil samples with their dates mean anything.

Questions:

1. What is the importance of Archaeopteryx to the evolutionary story?

2. What do the various dates assigned to the different candidates in the dino-to-bird story indicate?

3. What does the Bible teach about dino-to-bird evolution?

4. Apply the four powerful words to the view of archaeopteryx as an evolutionary link.

Credits and References

All charts, graphs, and images, unless otherwise noted, are created by Patrick Nurre, and used with his permission.

Clarifying the Conflict between Science and the Bible – Four Powerful Words
[1]"Billions and Billions of Demons" by Richard Lewontin, www.nybooks.com. January 9, 1997.
Exercise 2: [1]Explanations for the names of the geologic periods were drawn from these resources:
Cambrian - https://blogs.agu.org/mountainbeltway/files/2011/09/Geologic-Names-handout.pdf. Carboniferous - https://ucmp.berkeley.edu/carboniferous/carboniferous.php. Jurassic - https://blogs.agu.org/mountainbeltway/files/2011/09/Geologic-Names-handout.pdf
Cretaceous - http://www.geologypage.com/2014/04/cretaceous-period.html
Triassic –http://www.geologypage.com/2014/04/triassic-period.html.
Permian –https://www.britannica.com/science/Permian-Period.
Silurian and Ordovician –https://blogs.agu.org/mountainbeltway/files/2011/09/Geologic-Names-handout.pdf.
Devonian –https://blogs.agu.org/mountainbeltway/files/2011/09/Geologic-Names-handout.pdf.
[2]The Modern Geologic Time Scale or Time Column: Courtesy USGS.
Geologic Time Scale : Courtesy USGS, public domain
Exercise 3: [1] https://biologos.org/podcast-episodes/fossils [2] Howard, Gordon. "Index Fossils – Really?" *Creation* 34(4):52–55, October 2012, found at https://creation.com/index-fossils. [3]Woodmorappe, John. "The Geologic Column: Does it exist?" Found at https://creation.com/does-geologic-column-exist
Exercise 4: [1] From a picture by National Geographic Society, found at http://www.pbs.org/wgbh/evolution/library/03/4/image_pop/l_034_01.html. [2]Sloane, Christopher. *How Dinosaurs Took Flight: Fossils, Science, What we Think we Know, and Mysteries Yet Unsolved.* National Geographic Society, 2005, p.12-13.

Raised in beautiful Montana, Patrick Nurre spent many days rockhounding near the Big Horn River. This early interest led him to a lifelong study of the world of geology. Patrick was trained in secular geology, but after becoming a Christian, he began to seriously reevaluate his previous evolutionary beliefs, and became, instead, a young-earth creationist. Today, Patrick conducts geology classes (PreK-12) and seminars in the Seattle area, and speaks at numerous home school and church conventions on geology and our young earth. He also leads a variety of geology field trips every year, including Yellowstone National Park, where he helps families discover the Biblical geology of the Park. Patrick is the vision behind Northwest Treasures, a 501(c)3 corporation, devoted to creating excellent geology kits, curricula, and education for all ages. To help in this endeavor, Patrick has written 17 books on geology, for all ages. He is a long-standing participant in the IAC (International Association of Creation) Professional Development Program. In addition, he draws on over 40 years of pastoring experience. Patrick and his wife, Vicki, have three wonderful children and three grandchildren, and live in Bothell, Washington.

If you would like to contact Patrick about speaking or field trips: northwestexpedition@msn.com

For a list of speaking topics: NorthwestRockAndFossil.com

CPSIA information can be obtained
at www.ICGtesting.com
Printed in the USA
JSHW011928200423
40560JS00004B/15

9 780998 283296